THE ROLE OF SUPERCOMPUTERS IN SOLVING GLOBAL CHALLENGES BOOK

How High Performance Machines Are Redefining Global Cooperation

MARCUS T. HOOKS

COPYRIGHT

Copyright©2024 Marcus T. Hooks. All rights reserved. No part of this publication may be reproduced, distributed, or transmitted in any form or by any means, including photocopying, recording, or other electronic or mechanical methods, without the prior written permission of the publisher, except in the case of brief quotations embodied in critical reviews and certain other non-commercial uses permitted by copyright law

TABLE OF CONTENTS

COPYRIGHT 1

TABLE OF CONTENTS 2

INTRODUCTION 4

    He Role of Supercomputers in Solving Global Challenges Book 4

CHAPTER 1 13

    The Dawn of Supercomputing 13

CHAPTER 2 22

    Frontier: The First Exaflop Supercomputer 22

CHAPTER 3 32

    The Technology Behind the Power 32

CHAPTER 4 42

    Applications: Supercomputers in Science and Research 42

CHAPTER 5 52

    AI and Supercomputing: A Symbiotic Relationship 52

CHAPTER 6 61

    The Global Race for Supercomputing Supremacy 61

CHAPTER 7 70

Ethical Considerations and Risks of Supercomputing .. 70

CHAPTER 8 79

Supercomputing and the Environment: Powering the Future 79

CHAPTER 9 88

The Future of Supercomputing: Beyond the Exaflop Era 88

CONCLUSION 98

The Role of Supercomputers in Solving Global Challenges 98

INTRODUCTION

He Role of Supercomputers in Solving Global Challenges Book

In today's interconnected world, humanity faces an array of complex, multifaceted challenges: climate change, pandemics, food insecurity, energy shortages, geopolitical tensions, and many more. While these problems may seem overwhelming in their scope and urgency, the rapid advancement of technology provides us with unprecedented tools to confront them head-on. At the heart of this technological revolution is the supercomputer – an extraordinary machine capable of performing calculations at mind-boggling speeds, processing vast amounts of data, and running complex simulations that were once beyond the realm of possibility.

Supercomputers, with their ability to model and simulate intricate systems and predict outcomes with precision, have emerged as one of the most powerful forces in modern science, industry, and policy. These computing giants are not just advancing research but are actively contributing to solving some of the world's most pressing issues. From accelerating medical discoveries to optimizing energy systems, supercomputers are proving indispensable in the fight against global challenges.

Yet, despite their increasing role in global problem-solving, the full potential of these machines is often underestimated or misunderstood. Many people associate supercomputers with high-tech industries, military applications, or entertainment technology. However, their applications go far beyond these arenas. These machines have become instrumental in everything from mitigating the impacts of climate change and advancing renewable energy solutions to improving healthcare and addressing food shortages. The role of supercomputers in tackling these challenges cannot be overstated—they are driving innovations that shape the future of civilization.

This book is dedicated to exploring the pivotal role that supercomputers play in solving global challenges. It will delve into the specific ways in which supercomputers are transforming industries, enabling breakthroughs in research, and helping us anticipate and mitigate risks on a global scale. From understanding how these machines are built and operated to showcasing the real-world applications that are already making a significant impact, this book seeks to illuminate the extraordinary potential of supercomputing as a tool for positive change.

The Power of Supercomputing

At the heart of the supercomputing revolution lies one simple fact: computational power is growing exponentially. With each passing year, supercomputers grow faster, more efficient, and capable of processing exponentially larger datasets. As of 2024, the world's most powerful supercomputer, **Frontier**, based at Oak Ridge National Laboratory in the United States, can perform more than one exaflop—one quintillion (10^{18}) calculations per second. This achievement marks the beginning of the exaflop era, pushing the boundaries of what is computationally possible. The implications of this new capability are vast, and the areas it can impact seem almost limitless.

Supercomputers achieve their immense power by employing thousands, sometimes millions, of processors working in parallel. These parallel computing systems allow them to solve problems that are otherwise computationally prohibitive. This capability is essential for a range of applications—from simulating complex environmental systems and predicting natural disasters to advancing artificial intelligence and developing new medicines. The immense power of supercomputing lies in its ability to analyze massive datasets and perform simulations with precision and speed, unlocking insights that were once only imaginable.

Supercomputing and Climate Change

One of the most pressing global challenges today is climate change. The increasing frequency of extreme weather events, rising sea levels, and the loss of biodiversity have made it clear that urgent action is needed to mitigate the effects of global warming and limit its catastrophic impacts. Supercomputers are playing a critical role in this effort.

By simulating complex climate models, supercomputers help scientists predict future climate patterns, identify potential risks, and evaluate the effectiveness of various mitigation strategies. For example, supercomputers are used to predict how rising temperatures will affect weather patterns, agriculture, and ecosystems. These simulations provide valuable insights into how different regions of the world will be impacted by climate change, allowing policymakers to design more effective response strategies.

Moreover, supercomputers are essential in optimizing energy systems, particularly in the realm of renewable energy. With global demand for clean energy rising, supercomputers are being used to model new energy systems, improve energy storage, and design more efficient solar panels, wind turbines, and other green technologies. These technologies, once considered impractical or too costly, are now within reach due to the computational power provided by supercomputers.

Supercomputers in Healthcare and Medicine

The global health landscape is another area where supercomputers are proving indispensable. The COVID-19 pandemic demonstrated how crucial it is to have advanced computational tools for modeling the spread of infectious diseases, simulating the effects of interventions, and accelerating vaccine development. Supercomputers played a pivotal role in simulating virus behaviors, analyzing potential treatments, and aiding in drug discovery.

Beyond pandemics, supercomputers are being used in genomics and personalized medicine. By processing vast amounts of genetic data, supercomputers enable researchers to map out the human genome, understand genetic diseases, and develop personalized treatments. In the fight against cancer, for instance, supercomputers can simulate how different drug compounds might interact with specific cancer cells, allowing researchers to develop targeted therapies more efficiently.

Additionally, supercomputers play a role in advancing biomedical research by simulating protein folding and molecular interactions. This allows scientists to understand how diseases like Alzheimer's, Parkinson's, and other neurodegenerative disorders develop at a molecular level, leading to the development of more effective treatments.

Advancing Food Security and Agriculture

With a growing global population and climate change threatening agricultural yields, food security has become one of the most urgent issues facing the world today. Supercomputers are increasingly being used to address this challenge by modeling agricultural systems, optimizing crop yields, and predicting the impacts of climate change on food production.

For example, supercomputers can model soil conditions, weather patterns, and crop growth to help farmers make data-driven decisions about irrigation, planting schedules, and pest management. These models can also simulate the effects of climate change on crop production, helping farmers prepare for shifts in growing conditions and improving food resilience.

Moreover, supercomputers are being used in the development of genetically modified crops that are more resistant to pests, drought, and disease. These innovations are crucial in ensuring that the world can feed a growing population while minimizing the environmental impact of agriculture.

Supercomputers in Geopolitics and National Security

In addition to their role in addressing global challenges like climate change and healthcare, supercomputers also play a critical role in geopolitics and national security. From cryptography to military simulations, these machines are essential tools for nations looking to maintain an edge in global competition.

Supercomputers are used to simulate military strategies, analyze geopolitical trends, and predict potential conflicts. They are also vital in cybersecurity, helping nations protect critical infrastructure and defend against cyber threats. Supercomputers can process vast amounts of data to detect patterns of malicious activity, identify vulnerabilities, and create more secure systems.

The Future of Supercomputing

Looking ahead, the future of supercomputing is nothing short of exciting. Quantum computing, artificial intelligence, and machine learning are rapidly advancing fields that could redefine what is possible with computational power. Supercomputers are already playing a role in these fields, but as quantum computers come online and AI models become even more sophisticated, the potential for innovation will skyrocket.

Furthermore, the integration of supercomputing with other cutting-edge technologies like 5G, the Internet of Things (IoT), and blockchain will likely open up new opportunities for solving global challenges in ways we can't yet fully imagine.

The role of supercomputers in solving global challenges is still in its early stages, but their impact is already profound. From climate modeling and renewable energy solutions to healthcare and food security, these machines are transforming industries and enabling breakthroughs that were once out of reach. As technology continues to evolve, the capabilities of supercomputers will only expand, making them an even more powerful tool for addressing the world's most pressing problems.

This book aims to provide readers with a deeper understanding of how supercomputers work, the challenges they help solve, and the transformative potential they hold for the future. The next chapters will take a closer look at the specific applications of supercomputers in key areas such as climate change, healthcare, agriculture, and beyond. By the end of this book, readers will gain a greater appreciation for the role these machines play in shaping a better world and the vital contributions they make to solving some of humanity's most urgent problems.

CHAPTER 1
The Dawn of Supercomputing

The history of supercomputing is a story of continuous innovation, extraordinary vision, and relentless pursuit of computational power. From the humble beginnings of the earliest machines to the monumental leaps made in the 21st century, the evolution of supercomputers is a testament to human ingenuity and the desire to solve increasingly complex problems. Today, we stand at the threshold of a new era in computing—one in which supercomputers are capable of processing mind-boggling amounts of data, running simulations that can model entire ecosystems, and even predicting the future of climate change. To understand how we arrived at the present, it is essential to explore the history of supercomputing, the visionaries who made it possible, and the technological milestones that brought us into the exaflop era.

Early Supercomputers: The Beginnings of High-Performance Computing

The term "supercomputer" was first coined in the 1960s, but its roots go back even further, to the early days of computing in the 1940s and 1950s. In those early years, computers were primarily used for basic calculations and military purposes, such as code-breaking during World War II. Machines like the **ENIAC** (Electronic Numerical Integrator and Computer), which was completed in 1945, represented some of the first real breakthroughs in computing. ENIAC was capable of performing 5,000 basic calculations per second, a speed that was unimaginable at the time.

However, ENIAC and its successors were not considered "supercomputers" in the modern sense. They were room-sized machines with limited functionality and high maintenance costs. The first true supercomputers were designed with the purpose of solving specialized scientific and engineering problems that required massive computational power.

One of the first pioneers of supercomputing was **Seymour Cray**, often referred to as the "father of supercomputing." In the 1950s and 1960s, Cray was working with computers that were exponentially faster than any machines available at the time. As a young engineer at **Control Data Corporation (CDC)**, he helped design the **CDC 1604**, one of the first computers capable of running real-time applications. However, Cray's real legacy began with his design of the

CDC 6600, released in 1964, which was the world's fastest computer at the time.

The CDC 6600, with its **3 megaflops** (three million floating-point operations per second), marked the beginning of the era of supercomputers. Its performance was revolutionary, capable of handling complex scientific simulations far beyond the capabilities of its predecessors. This machine laid the groundwork for future supercomputing and established Cray as a central figure in the field.

In 1976, Seymour Cray left CDC to form his own company, **Cray Research**, and it was here that the next landmark in supercomputing history was created. The **Cray-1**, released in 1976, was the first supercomputer designed specifically for scientific applications. With a peak performance of **80 megaflops**, the Cray-1 was not only vastly more powerful than its contemporaries but also featured a revolutionary design that focused on high-speed data access and parallel processing. The Cray-1's sleek, curved design became iconic and symbolized the power of supercomputing.

The Evolution of Parallel Processing

While early supercomputers like the CDC 6600 and the Cray-1 focused on increasing processing speed through faster individual processors, the next step in the evolution of supercomputing came with parallel processing. Parallel processing involves breaking down complex tasks into smaller sub-tasks that can be processed simultaneously, vastly increasing computational speed.

In the 1980s and 1990s, the concept of parallel computing began to gain traction as researchers and engineers sought ways to further increase the computational power of supercomputers. The challenge, however, was to design systems that could effectively manage multiple processors working together in parallel without creating bottlenecks or inefficiencies.

A key milestone in this effort was the development of the **Connection Machine** in the mid-1980s by **Danny Hillis** and his team at Thinking Machines Corporation. The Connection Machine used a massively parallel architecture, with up to 65,536 processors working in tandem, allowing it to perform complex computations that were impossible for traditional, single-processor machines. While the Connection Machine was never commercially successful, its design philosophy had a profound influence on future supercomputers, demonstrating the potential of parallel architectures.

Parallel processing would continue to evolve over the following decades, leading to the development of more sophisticated and scalable systems. By the late 1990s, companies like **IBM**, **Intel**, and **Silicon Graphics** had begun to introduce new supercomputing systems based on massively parallel processing (MPP) architectures, which allowed supercomputers to scale up and tackle larger, more complex problems.

One of the most significant breakthroughs in this era was the introduction of **cluster computing**, which involved connecting multiple smaller computers (or nodes) together to create a powerful network. This approach allowed researchers to build supercomputers by linking together off-the-shelf components, dramatically lowering the cost of building high-performance systems while still providing impressive computational power.

The Growth of Computing Power: From Gigaflops to Petaflops

The 1990s and 2000s saw rapid advancements in both hardware and software, which spurred the growth of computing power. During this period, supercomputers reached a new milestone with the advent of **gigaflops** (billions of floating-point operations per second), **teraflops** (trillions of operations per second), and eventually **petaflops** (quadrillions of operations per second).

In 1997, the **ASCI Red** supercomputer, built by **Intel** for the **Los Alamos National Laboratory**, became the world's first supercomputer to achieve a sustained performance of **1 teraflop**. ASCI Red was used primarily for nuclear simulations, helping the U.S. government maintain the reliability of its nuclear stockpile without the need for nuclear testing.

This achievement was followed by several other milestones, including the development of the **IBM Blue Gene** series, which broke records for computational power. By 2008, **Blue Gene/P** had achieved **1 petaflop**, a monumental achievement that represented a major leap in the evolution of supercomputing.

The introduction of **parallel processing** and **distributed computing** allowed supercomputers to break new records, with machines able to solve ever more complex problems in physics, engineering, climate science, and medicine. The growth of the internet and the increasing amount of data being generated worldwide also played a critical role in driving the need for more powerful supercomputers, as researchers sought ways to analyze and understand vast datasets in real time.

Entering the Exaflop Era

While the petaflop era marked significant progress in computing power, the real leap came with the arrival of **exascale computing**. **Exascale computing**, or the ability to perform **1 exaflop** (1 quintillion operations per second), is considered the next frontier in high-performance computing. The first supercomputer to reach exaflop performance was **Frontier**, a machine developed by **Hewlett Packard Enterprise (HPE)** and installed at **Oak Ridge National Laboratory (ORNL)** in Tennessee.

In 2022, Frontier officially broke the exaflop barrier, achieving a peak performance of **1.1 exaflops**, making it the world's most powerful supercomputer. Frontier's capability represents a milestone that has vast implications for scientific research, artificial intelligence, and the ability to model highly complex systems in real time.

The journey from megaflops to exaflops has been a remarkable one, driven by advances in processor technology, memory architecture, and interconnects that allow supercomputers to scale up in both size and speed. The success of Frontier and other exascale systems is not only a technological achievement but also a critical tool for addressing some of the world's most urgent problems, from climate modeling to drug discovery.

The Future of Supercomputing

The development of supercomputers from their early days to the present reflects an ongoing quest for greater computational power. Pioneers like Seymour Cray and institutions like Los Alamos and Oak Ridge National Laboratory have been instrumental in pushing the boundaries of what's possible in computing. As we enter the exaflop era, supercomputers are poised to tackle problems that were once thought impossible.

Supercomputing will continue to evolve, with innovations such as quantum computing on the horizon, further enhancing the potential of these machines. As they become increasingly capable, supercomputers will play a central role in solving global challenges—addressing everything from climate change to global health crises. As we look to the future, we can expect that the incredible power of these machines will only grow, unlocking new possibilities and solutions for a rapidly changing world.

In the following chapters, we will explore how supercomputers are already being used to confront some of the world's most pressing issues and examine their potential to shape the future in ways we have yet to fully imagine.

CHAPTER 2

Frontier: The First Exaflop Supercomputer

The **Frontier supercomputer** is a landmark achievement in the world of high-performance computing (HPC). As the world's first supercomputer to reach **exaflop-level** performance, Frontier represents a leap forward in computational power that will drive a new era of scientific discovery and innovation. Developed by **Oak Ridge National Laboratory (ORNL)** in Tennessee, Frontier was designed to tackle the world's most complex scientific and engineering challenges, from climate modeling to drug discovery and artificial intelligence (AI). With its unparalleled capabilities, Frontier holds the key to advancing fields that require massive computational resources, enabling researchers to model, simulate, and analyze data on an unprecedented scale.

In this chapter, we will explore the Frontier supercomputer's design, architecture, key features, performance benchmarks, and the significance of its achievement in surpassing the **exaflop barrier** — a threshold that was once thought to be a distant aspiration in the world of supercomputing. By understanding Frontier's innovative design and immense power, we can gain a clearer picture of its role in shaping the future of computational science.

The Architecture of Frontier: A Fusion of Cutting-Edge Technologies

Frontier is built on a unique and highly specialized architecture that combines powerful **CPU** (Central Processing Unit) and **GPU** (Graphics Processing Unit) technologies, creating a hybrid system capable of handling the most demanding workloads across various scientific fields. The design is based on a **heterogeneous computing model**, where CPUs and GPUs work together to process different types of computational tasks simultaneously. This design helps achieve maximum performance efficiency for diverse applications, from traditional scientific simulations to AI and machine learning workloads.

1. CPUs and AMD's EPYC Processors

At the heart of Frontier's architecture are AMD's **EPYC processors**. These processors are based on the company's **Zen 4 microarchitecture**, which provides exceptional parallel processing capabilities. Each node in Frontier's system is equipped with **AMD EPYC 64-core processors**, providing an immense amount of processing power. These

CPUs handle traditional compute-intensive tasks such as simulations, data analysis, and scientific computations that do not require the specialized processing power of GPUs.

The **Zen 4 architecture** provides higher performance per watt and greater scalability than previous generations, making it a crucial element in enabling Frontier to achieve exaflop-level performance. EPYC processors are designed to handle large amounts of data throughput and support advanced memory management, both critical in addressing the increasing complexity of modern scientific problems.

2. GPUs and AMD's Instinct MI200 Accelerators

One of the most innovative aspects of Frontier's design is its integration of **AMD Instinct MI200 GPUs**, which provide massive parallel processing capabilities and are optimized for tasks such as **AI training**, **deep learning**, and **scientific simulations**. The MI200 GPUs, based on the **CDNA 2 architecture**, are specifically designed to perform at extreme levels of performance and efficiency, making them ideal for the types of workloads Frontier is expected to handle.

Each node in Frontier is equipped with four MI200 GPUs, with the **MI250X** model being one of the most powerful processors in the series. These GPUs are capable of performing complex calculations at unprecedented speeds, contributing heavily to Frontier's ability to break the exaflop barrier. By providing 10x the performance of traditional CPUs, the GPUs allow Frontier to process large datasets much faster than conventional computing systems.

3. The Slingshot Interconnect Network

The Frontier supercomputer is connected using a high-speed **Slingshot interconnect network**, which facilitates communication between thousands of nodes with minimal latency and maximum bandwidth. Developed by **Cray**, Slingshot ensures that the massive amount of data being processed by the supercomputer flows seamlessly between processors and memory systems, reducing bottlenecks and allowing for more efficient parallel computing.

The interconnect technology is critical to Frontier's performance because it enables the rapid transmission of data across its sprawling architecture, which consists of tens of thousands of nodes. Slingshot allows Frontier to scale efficiently, ensuring that the system can handle a wide range of scientific applications, from simulating molecular interactions to modeling the behavior of galaxies.

Performance Benchmarks: Reaching the Exaflop Barrier

The term **exaflop** refers to the ability to perform **one quintillion (10^{18})** floating-point operations per second (FLOPS). Achieving this level of performance was a long-standing goal in supercomputing, as it represents an order of magnitude greater than the previous benchmark of **petaflop** (one quadrillion FLOPS). The achievement of exaflop-level performance marks a monumental milestone for computational science, pushing the boundaries of what's possible in simulation, modeling, and data analysis.

Frontier's performance benchmarks are nothing short of extraordinary. According to the **TOP500** list of the world's fastest supercomputers, Frontier surpassed the exaflop barrier with a peak performance of **1.1 exaflops**. This makes it the world's fastest supercomputer, outperforming its closest competitors by a wide margin. The performance is measured using two primary benchmarks: **HPL (High-Performance Linpack)** and **HPCG (High-Performance Conjugate Gradient)**.

1. HPL Benchmarking

The **HPL benchmark** measures the speed at which a supercomputer can solve a system of linear equations, which is a fundamental task in scientific computing. Frontier's performance on the HPL benchmark exceeded the 1 exaflop mark, demonstrating its capacity for handling large-scale computations. In practice, this level of performance means that Frontier can solve complex problems that would take years on traditional computing systems in a fraction of the time.

2. HPCG Benchmarking

The **HPCG benchmark** evaluates a system's performance in solving sparse linear systems, a task common in simulations of physical systems like fluid dynamics, climate modeling, and nuclear simulations. Frontier also excelled in this benchmark, demonstrating its ability to tackle more realistic, real-world scientific problems that go beyond theoretical peak performance.

Together, these benchmarks underscore Frontier's ability to not only perform raw calculations at extraordinary speeds but also to efficiently handle the types of computational tasks that are most important for advancing science and technology.

Key Features and Significance of Frontier

1. Revolutionary Computational Power

Frontier's sheer computational power enables it to handle workloads that were once thought to be beyond the reach of even the most advanced supercomputers. This includes **large-scale simulations** of complex systems such as the human brain, advanced climate models, and even the behavior of quantum systems. Frontier's power also paves the way for breakthroughs in **AI research** by enabling the training of machine learning models that require immense amounts of data and processing power.

For instance, AI models that involve deep neural networks can be trained on massive datasets, improving their performance and enabling new AI applications in fields like drug discovery, autonomous vehicles, and natural language processing. The integration of both CPUs and GPUs allows for a more efficient use of computational resources, optimizing Frontier's performance across different domains of research.

2. Enhanced Scientific Discovery

Frontier is poised to revolutionize scientific discovery by enabling simulations that were previously too resource-intensive to carry out. For example, it allows scientists to simulate the behavior of molecules in a way that could lead to breakthroughs in drug development and material science. This has the potential to dramatically speed up the discovery of new medicines and technologies, ultimately improving quality of life.

In the field of **climate science**, Frontier can model the interactions of atmospheric, oceanic, and land-based systems in unprecedented detail. This level of simulation is critical in understanding climate change, predicting weather patterns, and developing strategies for mitigating the effects of global warming.

3. Support for National Security and Defense

Beyond its scientific applications, Frontier plays a key role in supporting **national security** and defense. The U.S. Department of Energy (DOE) uses Frontier to conduct simulations and analyses that help inform national defense strategies. This includes modeling the behavior of nuclear materials, simulating the effects of various weapons systems, and developing strategies to protect against cyber threats.

4. Exaflop as the New Frontier

Reaching the exaflop barrier is not just a technical achievement; it represents the future of supercomputing. The success of Frontier paves the way for other exaflop systems to emerge, each contributing to different scientific and industrial domains. As more organizations and institutions adopt exaflop-class supercomputers, the potential for solving some of humanity's most pressing challenges will continue to expand.

A Game-Changer for Computational Power

Frontier is a landmark in the evolution of supercomputing. Its design, combining powerful CPUs, GPUs, and high-performance interconnects, enables it to achieve extraordinary levels of computational power. The ability to break the exaflop barrier opens up new possibilities in **scientific research**, **AI development**, **climate modeling**, and **national defense**, making Frontier a true game-changer in the world of high-performance computing.

As we look to the future, systems like Frontier will continue to push the limits of what is possible in computational science, helping humanity tackle the most complex and pressing challenges of our time. The achievement of exaflop performance is not an end, but rather the beginning of a new chapter in the story of supercomputing. With Frontier leading the charge, the world stands on the cusp of an era where simulations, discoveries, and innovations will be realized faster and more efficiently than ever before.

CHAPTER 3

The Technology Behind the Power

Supercomputers like **Frontier** have revolutionized the way we approach computational problems. Their immense processing power has made previously unimaginable feats of scientific discovery possible. However, the achievement of such computational power doesn't come from a single breakthrough but from a complex interplay of cutting-edge technologies, including **processors, graphics processing units (GPUs)**, specialized hardware, **energy-efficient systems**, **cooling technologies**, and **advancements in chip design**.

Understanding the technological underpinnings of supercomputers requires a deep dive into how these systems are built, how they operate, and how the innovations behind them allow them to achieve extraordinary performance levels. In this chapter, we'll examine these critical components in detail to understand the immense power that drives systems like Frontier, focusing on the hardware and energy-efficient techniques that make such systems possible.

High-Performance Computing Hardware: The Backbone of Supercomputing

At the core of any supercomputer is its **hardware**, a combination of processors, memory, storage, and networking components that must work together seamlessly to process vast amounts of data at incredible speeds. The hardware infrastructure for supercomputers like Frontier must handle everything from the most complex simulations to real-time data processing across distributed systems.

1. Processors: The Heart of a Supercomputer

Supercomputers like Frontier rely on a combination of **Central Processing Units (CPUs)** and **Graphics Processing Units (GPUs)** to deliver unparalleled computational power. These processors are specifically designed to handle vast amounts of data in parallel, a key feature that distinguishes supercomputers from more conventional computers. While **CPUs** are optimized for sequential processing tasks and general-purpose computing, **GPUs** are designed to process many tasks simultaneously, making them ideal for high-performance computing (HPC) tasks, machine learning, and scientific simulations.

- **CPUs**: Frontier uses high-performance CPUs, such as the **AMD EPYC** processors, which are optimized for multi-threaded workloads. These CPUs are designed to process a large number of threads at once, which is essential for tasks like climate modeling, protein folding, and genomic sequencing. In Frontier, the AMD EPYC processors serve as the primary computational engines for a wide variety of scientific applications.

- **GPUs**: The **AMD Instinct MI250X GPUs** in Frontier take advantage of **SIMD (Single Instruction, Multiple Data)** and **SIMT (Single Instruction, Multiple Threads)** architectures to execute many tasks in parallel. These GPUs allow Frontier to accelerate workloads that require massive parallelism, such as deep learning and artificial intelligence (AI) computations. The use of GPUs helps Frontier reach exaflop performance, making it the world's first exaflop supercomputer.

GPUs and CPUs together enable Frontier to excel in diverse scientific fields. The **hybrid architecture** that combines these two types of processors allows for a **heterogeneous computing** environment where different components are optimized for specific tasks, enabling maximum computational efficiency.

2. Memory and Storage

Supercomputers require massive amounts of memory and storage to handle the vast datasets they process. In Frontier, the system utilizes **high-bandwidth memory (HBM)** and a custom-built **high-speed interconnect** to ensure that data flows seamlessly between processors and memory.

- **Memory**: Frontier uses **HBM2** (High Bandwidth Memory) technology, which allows data to be accessed at incredibly fast rates, ensuring that CPUs and GPUs can access the data they need without waiting for slower storage systems. This is crucial for workloads that require high-speed data access, such as scientific simulations and real-time weather forecasting.

- **Storage**: Frontier also features an advanced **data storage infrastructure** that allows for the rapid retrieval and processing of enormous datasets. The storage system is built using the latest **solid-state drive (SSD)** technology, which provides fast read and write speeds to support the massive data throughput required by the supercomputer.

3. Interconnects: The Glue That Binds Everything Together

The **interconnect** is the network system that links all of the components in a supercomputer. Frontier uses a high-speed interconnect called **Slingshot**, which is developed by **Cray**, a subsidiary of Hewlett Packard Enterprise (HPE). Slingshot is a **high-bandwidth, low-latency interconnect** designed to handle the massive data transfer requirements of supercomputing systems.

In Frontier, the interconnect allows for seamless communication between the hundreds of thousands of CPUs and GPUs. It ensures that data can move quickly across the system, allowing Frontier to process vast amounts of data in parallel, which is essential for achieving exaflop performance.

Energy Efficiency: Reducing the Environmental Impact of Supercomputing

While the performance of supercomputers like Frontier is staggering, their energy consumption is also a critical concern. As computational power increases, so does the demand for electricity, and supercomputing facilities often face significant challenges in managing power usage without compromising performance.

Supercomputing systems like Frontier are built with energy efficiency in mind. The development of **energy-efficient technologies** has become increasingly important in the race to push computational boundaries while minimizing the environmental impact of these systems.

1. Energy-Efficient Hardware

A key factor in Frontier's efficiency is its energy-efficient hardware. The system is designed to operate at peak performance without excessive power consumption. Frontier uses custom-designed components, such as **AMD EPYC processors** and **AMD Instinct MI250X GPUs**, which are optimized for performance per watt. These components are engineered to deliver the best possible performance while minimizing power usage, a critical factor when building systems that consume enormous amounts of energy.

2. Power Management and Cooling Technologies

Even the most energy-efficient hardware generates heat, and supercomputing systems need effective cooling solutions to ensure that temperatures remain manageable. Frontier employs **advanced cooling technologies** to keep its components within optimal operating temperatures.

- **Liquid Cooling**: Frontier utilizes **liquid cooling** technology to dissipate the heat generated by its processors and GPUs. This system circulates a coolant fluid around the components to absorb and carry away heat. Liquid cooling is more efficient than traditional air cooling, as water has a higher heat capacity than air, allowing for faster heat dissipation.

- **Direct-to-Chip Cooling**: Frontier's cooling system also includes **direct-to-chip cooling**, a technique where the coolant is brought into direct contact with the chips themselves. This ensures that heat is removed at the source, preventing overheating and maintaining performance efficiency.

3. Innovative Power Management Systems

In addition to cooling technologies, Frontier includes **power management systems** that monitor and optimize the energy usage of its hardware. These systems can adjust the power supply to different components based on their current workload, ensuring that the system is not consuming more energy than necessary.

For example, Frontier uses dynamic voltage and frequency scaling (DVFS) to adjust the power delivered to processors and GPUs depending on the computational demands of the task at hand. This allows the system to scale its power usage in real-time, reducing energy consumption during periods of low activity.

Advancements in Chip Design: Paving the Way for Future Supercomputing

As computing demands increase, chip design must evolve to meet these challenges. The processors and GPUs in supercomputers like Frontier are the result of **decades of research** and **innovation** in semiconductor technology. Advances in chip design are central to the development of high-performance computing systems that can meet the demands of modern applications.

1. 3D Chip Stacking

One of the key advancements in chip design is **3D chip stacking**. This technique allows multiple layers of chips to be stacked on top of each other, creating more compact, efficient processors. In Frontier, 3D stacking technology is used in combination with **high-bandwidth memory (HBM)** to reduce latency and increase data throughput. By stacking memory chips directly on top of the processor, data can be transferred much more quickly, enhancing overall system performance.

2. Custom Processors for Specific Workloads

Another trend in chip design is the development of **custom processors** tailored to specific workloads. While general-purpose CPUs and GPUs can handle a wide range of tasks, specialized processors are designed to optimize performance for particular applications. In Frontier, custom processors are used for tasks such as **machine learning** and **scientific simulations**, enabling the system to achieve high performance for a diverse set of workloads.

For example, **AI accelerators** like **tensor processing units (TPUs)**, which are designed specifically for deep learning tasks, can be integrated into supercomputers to accelerate AI and machine learning workflows. These specialized processors are an essential part of Frontier's ability to handle modern workloads such as AI model training and simulation.

3. Advanced Fabrication Processes

The progress in **chip fabrication** technology also plays a crucial role in the development of supercomputers like Frontier. Modern chips are manufactured using smaller **process nodes**, typically measured in **nanometers (nm)**. Frontier's processors are based on **7nm and 5nm fabrication processes**, which allow for more transistors to be packed into a given area, resulting in faster, more powerful chips.

These smaller process nodes allow for improved power efficiency, as the smaller size of the transistors reduces the amount of power required to switch them on and off. This, in turn, leads to faster performance and reduced energy consumption, crucial aspects of Frontier's design.

The Cutting Edge of Computational Power

Supercomputers like Frontier represent the pinnacle of what is possible with modern computing hardware. From processors and GPUs to specialized hardware and cooling technologies, every aspect of Frontier's design has been engineered for maximum performance and energy efficiency. These technologies are not just revolutionary for supercomputing—they are shaping the future of scientific research, healthcare, artificial intelligence, climate modeling, and beyond.

CHAPTER 4

Applications: Supercomputers in Science and Research

Supercomputers have become indispensable tools in the modern research landscape. Their immense computational power allows scientists and researchers to solve problems that were once deemed impossible. Fields such as **climate modeling**, **genomics**, **material science**, and **pharmaceuticals** are all being transformed by the capabilities of supercomputing. The ability to process vast amounts of data in real time, run highly complex simulations, and model systems at previously unimaginable scales has unlocked new realms of discovery.

One of the most groundbreaking supercomputers to contribute to these advancements is **Frontier**, developed at **Oak Ridge National Laboratory (ORNL)**. Frontier's exaflop-level performance—capable of performing one quintillion calculations per second—makes it a game-changer in research, allowing researchers to accelerate scientific breakthroughs and push the boundaries of knowledge. In this chapter, we will explore how supercomputers are making significant contributions to various scientific domains, with specific case studies showcasing their impact. We will also highlight how **Frontier** is playing a pivotal role in accelerating the pace of discovery.

Supercomputers and Climate Modeling

Climate modeling is one of the most critical areas of scientific research in the 21st century. As climate change accelerates, there is an urgent need to understand the complexities of the Earth's climate system and predict future conditions. Supercomputers, such as Frontier, are playing a central role in this process, enabling researchers to build more accurate and detailed models of the Earth's atmosphere, oceans, and land surfaces.

The Role of Supercomputing in Climate Prediction

In climate modeling, researchers simulate various climate scenarios, such as global warming, extreme weather events, and the effects of greenhouse gases. These simulations require enormous computational resources due to the large volumes of data involved and the complexity of the underlying physical processes. Supercomputers provide the necessary computational power to run models at a global scale and with higher resolution than was ever previously possible.

One of the most notable examples of supercomputers being used in climate modeling is the **Community Earth System Model (CESM)**, which has been used to simulate climate scenarios over centuries or even millennia. On systems like Frontier, these models can be run at higher resolutions, meaning scientists can simulate climate interactions in greater detail, leading to more accurate predictions about future conditions.

Case Study: Frontier's Contribution to Climate Science

Frontier's exaflop capability has already proven to be an invaluable tool in the field of climate science. It allows researchers to simulate and analyze climate systems with unprecedented detail and accuracy. For example, Frontier has been used to study the impact of melting ice sheets on global sea levels. By processing vast amounts of data from satellite observations, Frontier helps researchers understand the underlying dynamics of ice sheet movements and their potential impact on future climate scenarios.

Moreover, Frontier's computing power enables simulations of more complex atmospheric phenomena, such as hurricanes and thunderstorms, which require large-scale computations to capture the interactions of weather patterns in real time. These simulations help meteorologists predict the severity and potential impact of such events with a greater degree of confidence.

Supercomputers in Genomics: Accelerating the Study of DNA

The field of **genomics** has made remarkable strides in recent years, particularly with the advent of next-generation sequencing (NGS) technologies that enable scientists to sequence entire genomes in a matter of hours. However, analyzing the massive datasets generated by these technologies requires computational power that exceeds the capabilities of standard computing systems. Supercomputers like Frontier are now essential tools in the study of genomics, enabling researchers to process and analyze genomic data at scales that were once unthinkable.

Genomic Sequencing and Data Processing

Genomic sequencing generates vast amounts of data that must be processed and analyzed to identify genetic variants, disease markers, and other biological insights. For example, the **Human Genome Project**, which aimed to sequence the entire human genome, generated over 200 gigabytes of data—a small fraction of the data generated by modern sequencing technologies. With the explosion of genomic data, researchers are turning to supercomputers to analyze this information efficiently.

Supercomputers play a vital role in **genomic data processing** by allowing researchers to perform tasks such as **alignment** (matching DNA sequences to a reference genome), **variant calling** (identifying genetic variations), and **annotation** (identifying the functional elements in a genome). The ability to process genomic data at scale has accelerated discoveries in disease research, drug development, and personalized medicine.

Case Study: Frontier in Genomic Research

Frontier has been instrumental in advancing research in **personalized medicine** and **genetic disease research**. One of the groundbreaking projects enabled by supercomputing is the study of the **genetic basis of cancer**. Researchers are using supercomputers like Frontier to analyze vast datasets of DNA sequences from cancer patients, identifying genetic mutations that may contribute to tumor development. With Frontier's power, researchers can now examine thousands of genomes in parallel, providing more comprehensive insights into the genetic causes of cancer and other diseases.

Additionally, supercomputers enable **gene expression analysis** by processing data from **RNA sequencing** experiments, which measure the activity of genes under different conditions. This information is crucial for understanding how genes are regulated and how their expression is linked to diseases like Alzheimer's, Parkinson's, and cardiovascular disorders. Frontier's capacity to handle large-scale genomic data allows for faster identification of disease-related genes and the development of targeted therapies.

Supercomputers and Material Science: Accelerating Innovation in Engineering

The field of **material science** involves the study of the properties and behaviors of materials, with the goal of designing new substances that can be used in a wide range of applications, from manufacturing to energy storage. Supercomputers play a crucial role in simulating the behavior of materials at the atomic and molecular levels, providing insights that can lead to the development of more efficient, durable, and sustainable materials.

Material Simulation and Design

Material science has traditionally relied on experimental methods to explore the properties of new materials, which can be time-consuming and costly. Supercomputers have revolutionized this process by enabling **computer-aided** design (CAD) and **molecular dynamics simulations** that model the behavior of materials under various conditions. These simulations allow scientists to predict the properties of new materials before they are physically created in a lab.

The ability to simulate the interactions between atoms and molecules at the quantum level has opened up new possibilities for designing materials with specific properties, such as higher strength, better conductivity, or greater resistance to heat. Supercomputers like Frontier enable scientists to perform **first-principles calculations** based on quantum mechanics, providing accurate predictions about the behavior of materials at a fundamental level.

Case Study: Frontier's Role in Material Science

In material science, one key area where Frontier is making an impact is in the development of **new energy materials**, such as those used in **solar cells** and **batteries**. By simulating the behavior of materials at the atomic scale, Frontier allows researchers to identify new materials that could improve the efficiency of solar panels or extend the lifespan of batteries. This research is crucial for advancing renewable energy technologies and reducing reliance on fossil fuels.

Frontier is also being used to explore the potential of **high-temperature superconductors**, which could revolutionize power transmission and magnetic levitation technologies. These materials, which exhibit zero electrical resistance at high temperatures, could drastically improve energy efficiency. Simulating the behavior of superconducting materials at high temperatures requires immense computational power, which is provided by Frontier's advanced capabilities.

Supercomputers in Pharmaceuticals: Speeding Up Drug Discovery

The pharmaceutical industry is another domain where supercomputing is making a profound impact. The process of drug discovery traditionally involves a long and expensive trial-and-error approach, in which researchers test thousands of compounds to identify those that have therapeutic potential. Supercomputers allow scientists to accelerate this process by using simulations to predict how drug molecules will interact with biological targets, thus streamlining the discovery of new medications.

Drug Discovery and Molecular Modeling

Drug discovery begins with the identification of a **biological target**, such as a protein or enzyme, that is involved in a disease. Supercomputers enable researchers to simulate how small molecules will interact with the target, providing insights into which compounds are likely to be effective as drugs. These simulations, known as **molecular docking** and **molecular dynamics simulations**, are computationally intensive tasks that require vast amounts of processing power. Supercomputers like Frontier provide the capacity to model and test thousands of potential drug candidates in a fraction of the time it would take using traditional methods.

Case Study: Frontier in Drug Discovery

Frontier has been involved in accelerating the discovery of new drugs for diseases such as **COVID-19**, **cancer**, and **neurodegenerative disorders**. In the case of COVID-19, researchers have used Frontier to simulate how various molecules, including existing drugs, interact with the SARS-CoV-2 virus. These simulations help identify promising candidates for further testing, potentially fast-tracking the development of antiviral drugs and vaccines.

In cancer research, Frontier has been used to simulate the interactions between potential drug compounds and **cancer-associated proteins**, helping to identify molecules that could disrupt the growth of tumors. By speeding up the process of identifying viable drug candidates, supercomputers like Frontier are contributing to the development of more effective treatments for cancer and other diseases.

Frontier's Role in Pushing the Boundaries of Knowledge

Supercomputers like Frontier are not just advancing the pace of research; they are fundamentally changing the way we approach scientific problems. From **climate modeling** and **genomics** to **material science** and **pharmaceuticals**, the computational power provided by supercomputers is unlocking new realms of discovery that were once beyond reach

CHAPTER 5

AI and Supercomputing: A Symbiotic Relationship

Artificial Intelligence (AI), **Machine Learning (ML)**, and **Deep Learning (DL)** have seen extraordinary growth in recent years, revolutionizing industries ranging from healthcare and finance to transportation and entertainment. At the heart of these advancements lies an essential partnership: the relationship between supercomputing and AI. Supercomputers like **Frontier**, the first exaflop machine, have become critical enablers of modern AI research and applications, providing the immense computational power needed to train large-scale models, process vast datasets, and accelerate the pace of innovation in the AI field.

Historically, AI research was constrained by limited computational resources. Early machine learning algorithms required only modest computing power, but as AI models became more complex—particularly in the areas of **deep learning** and **neural networks**—the computational demands grew exponentially. Training a state-of-the-art AI model today requires powerful machines capable of performing billions, if not trillions, of calculations per second. Supercomputers provide the necessary hardware, software, and processing power to meet these demands, enabling AI researchers to push the boundaries of what is possible.

In this chapter, we will explore the symbiotic relationship between AI and supercomputing. We will examine how supercomputers power AI research, contribute to the training of large-scale machine learning models, and unlock the potential for breakthroughs across various domains. Additionally, we will discuss the future of AI in relation to the ongoing advancements in supercomputing, considering the role these machines will play in shaping the next generation of AI systems.

Supercomputers and AI Research: Accelerating Innovation

AI research is no longer limited to small datasets or simple algorithms. Researchers are now working with massive datasets—ranging from medical images and genomic data to text, audio, and video—that require advanced techniques to analyze. This is where supercomputing has become indispensable.

1. Training Large AI Models

Training AI models, especially deep learning models, requires substantial computational resources. The larger and more complex the model, the more calculations it needs to perform during the training phase. Models like **GPT (Generative Pretrained Transformer)**, which powers large language models such as **ChatGPT**, require not only enormous datasets but also vast computing power to train.

Supercomputers provide the necessary infrastructure to train these large models. Frontier, with its **exaflop-level performance**, enables researchers to run complex AI training workloads in record time. The process of training a deep learning model involves feeding it vast amounts of data, adjusting its internal parameters, and running multiple iterations to optimize its performance. Supercomputers allow for **parallel processing**, where multiple calculations are performed simultaneously, significantly reducing the time required to train a model.

For instance, AI models used in natural language processing (NLP), image recognition, or autonomous driving benefit immensely from supercomputing power. These models require enormous datasets, such as millions of hours of video footage, or billions of text samples, to learn patterns, make predictions, or recognize objects. Without the high-performance capabilities of supercomputers, training these models would be impractically slow or even impossible. Supercomputers can run multiple experiments at once, testing various algorithms, configurations, and optimizations to achieve the best results.

2. Accelerating AI Research Through Simulation

Beyond model training, supercomputers play a critical role in the **simulation** of AI systems in various real-world applications. Whether it's simulating how a **self-driving car** reacts to a specific set of conditions or testing how an AI agent performs in a virtual environment, supercomputers can simulate complex scenarios that would be difficult, time-consuming, or dangerous to conduct in real life.

These simulations are invaluable for improving AI systems, as they allow for fast iteration without the costs and risks of real-world testing. For example, AI models trained on **autonomous driving** technology use simulation to model thousands of potential driving scenarios—from weather conditions to traffic situations—to improve the model's decision-making abilities.

Supercomputers enable researchers to run these simulations at scale. The **parallel processing** capabilities of systems like

Frontier allow for the execution of millions of different driving scenarios at once, speeding up the research process and providing more comprehensive data for fine-tuning AI models.

3. AI for Scientific Discovery and Problem Solving

AI is also transforming scientific discovery. With the help of supercomputers, AI has been used in a variety of scientific fields to make breakthroughs in previously inaccessible areas. From **drug discovery** to **material science**, AI models are being trained on massive datasets to uncover new insights, discover patterns, and make predictions that were previously beyond human capability.

For example, AI algorithms powered by supercomputers are playing a significant role in identifying potential drug candidates. By analyzing vast chemical and biological datasets, AI can predict how different compounds will interact with specific biological targets. This speeds up the **drug discovery** process, which traditionally involves years of trial and error in the lab. Supercomputers provide the computational horsepower necessary to simulate the interactions between millions of compounds and their potential effects on human health.

In material science, AI models are being used to predict the properties of new materials before they are even synthesized. By feeding existing data into AI models, researchers can identify materials with specific properties, such as greater conductivity, improved strength, or enhanced flexibility, which can then be tested experimentally.

The Role of GPUs in AI and Supercomputing

One of the key technological advancements driving the AI revolution is the development of **Graphics Processing Units (GPUs)**. GPUs were originally designed for rendering graphics in video games, but their highly parallel architecture—capable of executing thousands of simple operations simultaneously—makes them ideal for AI and deep learning tasks. In fact, GPUs have become the preferred hardware for training deep learning models due to their ability to process large volumes of data quickly and efficiently.

Supercomputers like **Frontier** leverage GPUs to accelerate machine learning and deep learning tasks. In particular, **NVIDIA A100 GPUs** are used extensively in supercomputers to power AI workloads. These GPUs are designed to handle the immense computational requirements of AI, from processing large datasets to training complex neural networks.

The **hybrid computing architecture** of Frontier, which combines **CPUs** and **GPUs**, allows for a more balanced approach to AI and supercomputing. While CPUs handle tasks requiring high single-threaded performance, GPUs take on the heavy lifting of parallel processing. This collaboration enables researchers to harness the strengths of both processing units, improving overall efficiency and performance.

The Future of AI and Supercomputing: Scaling for the Next Generation

As AI research continues to evolve, supercomputing will play an even more crucial role in advancing the field. The demand for larger and more sophisticated models is expected to grow exponentially, and so too will the need for computational resources to support them. To meet these needs, supercomputers are continuously evolving.

1. Quantum Computing and AI

One of the most exciting developments in the world of supercomputing and AI is the rise of **quantum computing**. Quantum computers promise to solve certain types of problems much faster than classical computers by leveraging the principles of quantum mechanics. While quantum computing is still in its early stages, it holds the potential to revolutionize AI research.

Quantum computers could accelerate AI training by allowing for much faster processing of certain types of optimization problems. For example, training deep learning models often requires solving complex optimization tasks, which could be sped up significantly by quantum algorithms. As quantum computers become more powerful, they may work in tandem with classical supercomputers to provide an even greater boost to AI research.

2. AI for Supercomputing

The future of supercomputing is also linked to the development of AI itself. As supercomputers become more complex and capable, they will be increasingly reliant on AI to manage their operations, improve efficiency, and optimize performance. AI algorithms can be used to detect and address system inefficiencies, predict hardware failures, and automate the management of large-scale computing tasks.

Moreover, AI is poised to play a critical role in developing next-generation supercomputing hardware. Machine learning techniques are being used to design and optimize chip architectures, helping to build processors that are better suited to the demands of modern supercomputing workloads.

The Symbiosis of AI and Supercomputing

The relationship between AI and supercomputing is undeniably symbiotic. Supercomputers like Frontier provide the computational power needed to train, refine, and deploy advanced AI models, while AI enhances the capabilities of supercomputers by improving hardware efficiency, automating tasks, and advancing scientific discovery.

As AI continues to evolve and supercomputers become even more powerful, the boundaries of what is possible in both fields will expand. Supercomputers will continue to provide the computational foundation for breakthroughs in AI, and AI will drive innovations that push supercomputing capabilities to new heights. Together, they represent a transformative force that will shape the future of technology, science, and society.

CHAPTER 6
The Global Race for Supercomputing Supremacy

The pursuit of the world's most powerful supercomputer has become a high-stakes global competition, with countries and companies alike vying for technological supremacy. At the heart of this race is a desire to establish national leadership in science, innovation, and economic competitiveness. **Supercomputers** are no longer just tools for scientific discovery—they are critical assets in the geopolitical arena. The nation that controls the most powerful supercomputing systems gains not only an edge in research and development but also a significant advantage in national security, defense, and technological influence on the global stage.

The race for supercomputing supremacy is marked by a handful of key players—**Japan, the United States**, and **China**—all of which have made tremendous investments in building state-of-the-art machines capable of pushing the limits of computational performance. Nations have realized that supercomputing power is a strategic asset that can drive innovation across industries, strengthen military capabilities, and bolster economic competitiveness. As a result, supercomputing has become a central focus of both public and private sector initiatives, leading to intense competition to build the next "world's most powerful supercomputer."

In this chapter, we will explore the ongoing global race for supercomputing supremacy, focusing on key competitors such as **Fugaku** in Japan, **LUMI** in Finland, and other notable players. We will examine the geopolitical implications of supercomputing, particularly how these powerful machines are shaping national security, technological leadership, and global influence. Understanding this competition is essential not only for appreciating the role of supercomputing in scientific and technological development but also for grasping the broader political and economic forces driving innovation today.

Fugaku (Japan): Leading the Charge in Supercomputing

One of the key players in the global supercomputing race is **Fugaku**, a supercomputer developed by **Fujitsu** in collaboration with the **RIKEN** research institute in Japan. Fugaku made headlines in 2020 when it claimed the title of the world's most powerful supercomputer, surpassing China's **Sunway TaihuLight** and the United States' **Summit** (then held at Oak Ridge National Laboratory). Fugaku achieved a peak performance of **442 petaflops**, and later, after optimization and tuning, its performance reached an impressive **442 petaflops sustained** on the **High Performance Linpack (HPL)** benchmark, making it the world's fastest machine at the time.

The significance of Fugaku goes beyond its technical specifications. Japan's leadership in building Fugaku was driven by a number of strategic goals, including advancing the country's **AI**, **biotechnology**, **climate research**, and **materials science** capabilities. Fugaku's design is based on the **ARM architecture**, a departure from the traditional x86 architecture used by most supercomputers, marking a milestone in the diversification of computing architectures. This design allows for greater energy efficiency, which is a critical consideration in the growing scale of supercomputing, where power consumption has become one of the most pressing challenges.

Fugaku's impact on Japan's technological and scientific landscape has been profound. The machine has enabled researchers to simulate and analyze complex phenomena at unprecedented scales. Fugaku has been used in a variety of research areas, from **COVID-19** pandemic simulations to **climate change** modeling, **drug discovery**, and **materials science**. It also played a significant role in helping Japan's government understand the spread of the COVID-19 virus, assisting with the simulation of the dynamics of virus transmission and the effectiveness of interventions.

But Fugaku's geopolitical significance extends beyond its immediate contributions to science. In an era where technological dominance is closely linked to national security, Japan's investment in such a powerful supercomputer positions it as a key player in the global technological landscape. Fugaku gives Japan a critical advantage in the domains of **AI research**, **cybersecurity**, and **defense technologies**, aligning with the country's broader ambitions to strengthen its technological infrastructure and maintain its global leadership in scientific research.

LUMI (Finland): A European Contender

While Japan has made significant strides in the supercomputing race, Europe is also establishing itself as a major player. **LUMI**, a supercomputer located at the **CSC – IT Center for Science** in Finland, is one of the most powerful supercomputers in Europe. LUMI is a joint project between several European countries, and it stands as a prime example of Europe's commitment to advancing high-performance computing capabilities.

LUMI is part of the European **EuroHPC (European High-Performance Computing)** initiative, designed to create a network of supercomputing systems across Europe. LUMI, which officially came online in late 2021, has a peak performance of **375 petaflops** and is designed to be a central asset in the European Union's efforts to maintain technological sovereignty in the supercomputing domain. The machine is powered by **AMD EPYC** processors and **AMD Radeon Instinct** GPUs, reflecting Europe's push for energy-efficient, cutting-edge hardware.

The primary goal of LUMI is to serve as a resource for researchers across Europe, providing computational power to address challenges in **healthcare**, **climate science**, **energy**, **industrial applications**, and **AI research**. By providing access to such a powerful resource, LUMI aims to foster collaboration between European researchers and elevate Europe's status as a global leader in science and technology.

The geopolitical significance of LUMI is notable for its role in reducing Europe's dependency on supercomputing systems from non-European countries, especially the United States and China. By building and hosting LUMI on European soil, the European Union is asserting its independence in high-performance computing, ensuring that it has access to critical computational resources without relying on external powers. This is particularly important as the race for technological leadership becomes intertwined with issues of **data sovereignty**, **security**, and **economic competitiveness**.

Other Supercomputing Competitors: The United States and China

While Fugaku and LUMI represent significant strides in the race for supercomputing supremacy, the **United States** and **China** remain two of the most formidable competitors in the global race. The **United States** continues to invest heavily in supercomputing infrastructure, with projects such as **Frontier** at **Oak Ridge National Laboratory (ORNL)** and **Argonne National Laboratory's Aurora** system. As previously mentioned, Frontier is currently the fastest supercomputer in the world, capable of **1.1 exaflops**, a monumental achievement that far exceeds the performance of Fugaku.

China, too, remains a key player with several advanced supercomputing systems, such as **Tianhe-2** and **Sunway TaihuLight**. While China has faced challenges in terms of international access to the latest semiconductor technologies due to trade restrictions, it has nevertheless made significant strides in developing domestic supercomputing infrastructure. China's investments in **AI**, **quantum computing**, and **big data** are closely linked to the growth of its supercomputing capabilities, enabling the country to maintain its position as a global leader in technology.

Both the United States and China see supercomputing as essential to national security. The ability to simulate complex scenarios, conduct military research, and enhance intelligence capabilities relies heavily on access to cutting-edge supercomputing resources. In the context of AI and machine learning, both countries are engaged in an arms race to develop the most powerful AI systems, which in turn require the computational power provided by supercomputers.

The Role of Supercomputing in National Security and Technological Leadership

Supercomputing has increasingly become a critical tool for national security, with governments around the world recognizing its importance in defense, intelligence, and cybersecurity. The ability to simulate military operations, conduct cryptanalysis, and model complex scenarios in real time gives nations a strategic advantage in an era of technological warfare.

Supercomputers are used in **defense research** to simulate the behavior of weapons systems, predict the outcomes of military operations, and model geopolitical scenarios. For example, the U.S. Department of Defense relies on supercomputing systems to run simulations of missile defense systems, nuclear weapons testing, and advanced aircraft design. Similarly, supercomputing resources are critical in cyber defense, enabling rapid processing of data to detect and respond to cyber threats.

In addition to national security, the development of supercomputers is also linked to broader goals of **technological leadership** and **economic competitiveness**. Countries with the most powerful supercomputers have the ability to push the frontiers of scientific research, driving innovation in key sectors such as **artificial intelligence**, **drug discovery**, and **materials science**. Supercomputing gives these nations a competitive edge in attracting top talent, fostering collaboration, and securing global leadership in the digital economy.

The Geopolitical Implications of Supercomputing

The race for supercomputing supremacy is not just a competition of numbers and performance benchmarks—it is a battle for national prestige, security, and technological power. As countries invest heavily in developing the next generation of supercomputing systems, they are not just aiming to be the fastest—they are positioning themselves as leaders in the global digital economy.

The geopolitical implications of this race are far-reaching. Supercomputing has become a strategic asset in global power dynamics, influencing everything from national security and defense capabilities to international trade and diplomatic relationships. As the global supercomputing landscape continues to evolve, it will undoubtedly play a pivotal role in shaping the future of technology, science, and geopolitics.

CHAPTER 7
Ethical Considerations and Risks of Supercomputing

#The arrival of exaflop-level supercomputers, such as **Frontier**, has ushered in an unprecedented era of computational capability. These powerful machines enable advances in fields as diverse as climate modeling, genomics, and artificial intelligence (AI). However, the extraordinary power of supercomputing brings with it a host of **ethical concerns** and **risks** that need to be carefully considered. As supercomputing becomes increasingly central to innovation, it also raises profound questions about **privacy, security,** and **responsibility**.

Supercomputers are not only capable of processing vast amounts of scientific data; they also play a crucial role in increasingly sophisticated applications of AI, machine learning, and **autonomous systems**. These capabilities offer incredible opportunities but also present challenges that have the potential to affect society on a global scale. Issues such as **data privacy**, the role of supercomputers in **military** technologies, and the ethical implications of AI-driven decision-making are central to discussions on the ethical use of supercomputing.

In this chapter, we will examine the ethical risks associated with supercomputing, including the **use of supercomputers in surveillance**, **military applications**, and **AI decision-making**, as well as their potential unintended consequences. We will also delve into the responsibilities of those who create and use these technologies to ensure they are developed and applied in ways that are ethical, fair, and accountable.

The Ethical Implications of Supercomputing in Surveillance

One of the most immediate ethical concerns with the growing computational power of supercomputers is their potential use in **surveillance** systems. Supercomputers' ability to process and analyze massive datasets in real time has made them essential tools for governments and corporations in monitoring and tracking individuals. While such systems can be used to enhance **public safety**, they also carry significant risks to personal privacy and civil liberties.

Supercomputing power enables **advanced surveillance technologies** that can sift through vast amounts of data from social media, communications, financial transactions, and even personal location data. This capacity raises the possibility of **mass surveillance**, where individuals' movements, behaviors, and preferences are continuously tracked and analyzed without their knowledge or consent. Governments may use supercomputing to enhance **counterterrorism** efforts, **law enforcement** capabilities, and **national security**, but the same technologies can also be misused for political control or social manipulation.

Consider the case of China's **Social Credit System**, which uses AI and big data to monitor and score citizens based on their behavior, including their financial and social actions. The data required to operate such a system relies heavily on **supercomputing** to process millions of data points in real-time. While advocates claim that such surveillance programs can improve public order and ensure safety, critics argue that they represent a disturbing infringement on personal freedoms and privacy. These systems raise fundamental ethical questions about the **balance between security and individual rights** in a highly connected world.

The ethical risks of surveillance are compounded by the potential for **algorithmic bias** in AI systems used for data analysis. **AI decision-making** can perpetuate existing societal inequalities, especially when the data used to train

AI models reflects biases in society. For example, predictive policing algorithms—designed to forecast criminal activity—may disproportionately target marginalized communities if trained on biased historical data. Without careful attention to the ethical implications, surveillance technologies powered by supercomputers can exacerbate social injustices, further entrench inequality, and violate individuals' right to privacy.

Supercomputing and AI Decision-Making: The Ethical Dilemmas

As **artificial intelligence** (AI) and **machine learning** continue to evolve, the role of supercomputers in driving AI's capabilities has become more prominent. Supercomputers are not only enabling faster and more accurate training of AI models but also facilitating the creation of increasingly complex systems that can make decisions in real-time. From self-driving cars to automated hiring processes and predictive health diagnostics, AI-powered decision-making is already influencing lives on a global scale.

However, as AI systems become more autonomous, ethical questions arise regarding the **accountability** of decisions made by machines. **Supercomputers** have the capacity to process vast quantities of data, and AI algorithms powered by supercomputing can analyze patterns in ways that humans cannot. But when these systems make decisions that have a direct impact on people's lives—such as approving loans, determining criminal sentences, or diagnosing diseases—who is responsible if something goes wrong?

The use of supercomputing to enhance AI decision-making can result in significant ethical dilemmas. For example, **autonomous vehicles** powered by AI must make decisions in high-stress situations, such as choosing whom to harm in the event of an unavoidable crash. How should such decisions be made, and who is accountable for them? This issue has sparked debates about the role of **ethics in AI programming**, as well as the potential for bias and discrimination in decision-making processes.

One notorious example of algorithmic bias is the use of AI in **criminal justice**. Algorithms used to predict the likelihood of reoffending, such as the **Compas system** in the U.S., have been shown to disproportionately flag African American defendants as high-risk, despite no evidence that they are more likely to commit future crimes. The **ethical risks** of using AI in such high-stakes decisions are clear: AI systems can inadvertently reinforce societal biases and injustices if they are not carefully designed and scrutinized.

The role of supercomputing in **AI decision-making** also raises concerns about **transparency** and **explainability**. AI models, particularly those based on **deep learning**, can function as "black boxes," meaning their decision-making processes are often opaque to human observers. This lack of transparency can make it difficult to understand how decisions are made and whether those decisions are fair, ethical, or justifiable.

The Use of Supercomputing in Military Applications: Ethical Risks and Concerns

Another significant ethical concern surrounding supercomputing lies in its use for military purposes. **Supercomputers** have long been utilized in defense-related research, from simulating weapons systems to developing new technologies for national security. As supercomputing advances, its applications in **military** technology have grown increasingly sophisticated. Supercomputers are now being used for simulations of warfare scenarios, cryptography, **drone operations**, and autonomous weapons systems.

The use of AI in military applications powered by supercomputers presents a unique set of **ethical challenges**. **Autonomous weapons systems**, also known as "killer robots," are a particularly contentious area of concern. These systems are capable of making life-and-death decisions without human intervention, raising fears of accidental escalation, violations of international law, and the **moral implications** of allowing machines to make lethal decisions.

For example, **autonomous drones** can be used to target and eliminate individuals without direct human oversight, a practice that has already been employed by militaries around the world. These drones, equipped with AI and controlled by supercomputers, can identify targets, evaluate threats, and execute strikes in real time. While these technologies are seen as potentially reducing human casualties and increasing military efficiency, they also raise questions about **accountability** and **human rights**.

The ethical concerns around military AI are compounded by the prospect of **AI-powered cyber warfare**, in which supercomputers can be used to breach critical infrastructure, disrupt communication networks, and carry out surveillance operations. The use of supercomputing in cyber warfare represents a significant threat to global security and calls for international regulations to govern its use.

The geopolitical implications of supercomputing in military contexts are far-reaching. Nations that develop the most advanced supercomputing capabilities gain not only a technological edge in civilian applications but also the ability to exert influence and control in **military and defense** strategies. Supercomputers have thus become central to national security considerations, with governments investing heavily in developing next-generation systems that can provide advantages in warfare and defense technologies.

Balancing Innovation with Ethical Responsibility

As the power of supercomputers continues to grow, the ethical considerations surrounding their use become more critical. Supercomputing holds the potential to drive innovation, accelerate scientific discovery, and solve global challenges, but it also raises complex questions about privacy, accountability, and the role of AI in decision-making.

Ethical concerns regarding **surveillance**, **AI decision-making**, and **military applications** must be addressed proactively by both researchers and policymakers. As we develop and deploy more powerful technologies, it is essential that they are guided by strong ethical principles that prioritize human rights, transparency, and fairness. **International cooperation** and **regulation** will play key roles in ensuring that supercomputing advancements are used responsibly and that their benefits are harnessed for the greater good of humanity.

In the end, the challenge will be to balance the immense potential of supercomputing with the need to safeguard privacy, prevent misuse, and ensure that technological advancements are used ethically and for the benefit of society as a whole.

CHAPTER 8

Supercomputing and the Environment: Powering the Future

Supercomputers have revolutionized science, medicine, engineering, and countless other fields, enabling advancements that were previously unthinkable. However, as these systems grow more powerful, they come with significant **environmental costs**, particularly in terms of **energy consumption**. Supercomputers like **Frontier** and others at the cutting edge of high-performance computing (HPC) require massive amounts of electrical power to function. The **energy consumption** of modern supercomputers can be on the scale of entire cities, and this presents a pressing challenge for researchers and companies working in the field.

While the benefits of supercomputing are undeniable, from advancing climate modeling to developing life-saving drugs, the environmental impact of these machines cannot be ignored. As the demand for more powerful systems grows, so does the need for **sustainable solutions** that can mitigate the ecological consequences of supercomputing. There is an increasing urgency to make supercomputing **greener**, finding ways to reduce its energy consumption, improve its **efficiency**, and reduce its **carbon footprint**.

In this chapter, we will explore the environmental impact of supercomputing, focusing primarily on the **energy consumption** of these machines and the innovative steps being taken to address this issue. We will look at efforts to make supercomputing more **energy-efficient**, as well as advancements in **green computing** technologies that are helping to power the future of high-performance computing with sustainability in mind. Ultimately, the goal is to ensure that supercomputing can continue to drive progress without unduly harming the planet.

The Enormous Energy Consumption of Supercomputers

Supercomputers, by their very nature, require vast amounts of **electricity** to operate. A single supercomputer can consume anywhere from **several megawatts** to **tens of megawatts** of power, depending on its size and the intensity of its computational tasks. This energy consumption puts a significant strain on local electrical grids and contributes to a supercomputer's environmental footprint. To put this into perspective, the **Fugaku** supercomputer in Japan, once the world's most powerful, consumes around **30 megawatts** of electricity. **Frontier**, the first exaflop supercomputer in the

United States, is expected to have similarly high power requirements, making it one of the largest energy consumers in the world.

While this might seem like an unavoidable byproduct of the computational power that these systems deliver, the issue of energy consumption is more complex. The energy used by supercomputers primarily powers their **processors** (CPUs and GPUs), memory systems, and storage systems. However, the real energy burden often comes from the **cooling systems** required to prevent these machines from overheating due to the vast amount of heat generated during high-speed computations. These cooling systems can consume as much, if not more, energy than the supercomputer itself.

This reality raises several environmental and economic concerns, particularly when it comes to the sustainability of building and operating large-scale supercomputing systems. Supercomputers that rely on traditional cooling methods, such as air conditioning or **chilled water**, can quickly escalate the energy consumption of the facility. The growing energy demands of these systems are pushing the boundaries of the **energy grid** and contributing to the overall **carbon emissions** of data centers and supercomputing facilities.

Innovations in Green Computing: Reducing the Carbon Footprint

The increasing recognition of the environmental impact of supercomputing has spurred **innovation** in the field of **green computing**. Green computing refers to the practice of designing, developing, and operating computing systems in a manner that reduces energy consumption and minimizes environmental impact. Over the past decade, significant strides have been made in creating **energy-efficient hardware**, improving **cooling technologies**, and utilizing **renewable energy sources** to power supercomputers.

1. Energy-Efficient Hardware

One of the most direct ways to reduce the energy consumption of supercomputers is through the development of **energy-efficient hardware**. Advances in processor technology, such as **multi-core** and **multi-threaded** CPUs and GPUs, have enabled a dramatic increase in computational performance without a proportional increase in energy usage. For instance, **low-power chips** that focus on **parallel processing** can perform multiple operations simultaneously, reducing the overall energy consumption per calculation.

The shift from **traditional silicon-based chips** to more **specialized processing units** like **Graphics Processing Units (GPUs)** has also contributed to better energy efficiency. GPUs are optimized for high parallelism, meaning they can perform many computations at once, reducing the total time and energy needed for certain tasks. Innovations like **ARM-based processors** have also emerged as low-power alternatives to traditional x86 architectures, allowing for better energy optimization.

2. Optimized Cooling Solutions

Cooling is another major area of innovation aimed at reducing the environmental impact of supercomputing. Traditional air-conditioned cooling methods used in data centers are highly energy-intensive, often consuming as much electricity as the machines they are cooling. In response, researchers have developed several more **efficient cooling technologies** designed to lower energy usage while maintaining the optimal operational temperature of supercomputers.

One notable solution is **liquid cooling**, where supercomputers use **specialized coolants** instead of air to transfer heat away from the system. Liquid cooling is far more efficient than air cooling because liquids have a higher heat capacity and can carry heat away more effectively. In addition to traditional liquid cooling systems, more advanced solutions include **immersion cooling**, in which entire servers are submerged in non-conductive liquids that directly absorb the heat generated by the components.

Another innovative cooling technique is the use of **free air cooling** systems, which use naturally occurring cold air from the environment to cool down supercomputing systems, rather than relying on energy-intensive cooling machines. Facilities in **cold climates** can take advantage of this technology, significantly reducing the energy needed for cooling.

3. Renewable Energy Sources

The use of **renewable energy** sources, such as **solar, wind,** and **hydropower,** is perhaps the most promising way to make supercomputing more sustainable in the long run. As the demand for computing power grows, it becomes increasingly important to source the energy that powers these systems from sustainable sources.

For instance, some supercomputing centers, like those at **Oak Ridge National Laboratory,** are exploring partnerships with energy providers to utilize clean energy from solar farms, wind power, or hydropower to supply the electricity for their supercomputing operations. By using these renewable sources of power, supercomputing facilities can significantly reduce their **carbon footprint** and contribute to a more sustainable future.

Moreover, some supercomputing facilities are exploring the possibility of using **geothermal energy** or **direct air cooling** to power their operations in a way that minimizes their reliance on traditional, carbon-intensive power sources.

4. Energy-Aware Supercomputing Software

In addition to hardware and cooling innovations, software solutions also play a key role in reducing the energy consumption of supercomputing systems. Modern supercomputing environments incorporate **energy-aware software** that monitors and adjusts system performance based on energy consumption goals.

Energy-efficient scheduling algorithms allow supercomputing jobs to be assigned to resources in ways that maximize performance while minimizing energy usage. Additionally, **dynamic voltage and frequency scaling (DVFS)** technologies enable processors to adjust their power consumption in real time based on the workload, reducing energy use when full performance is not required.

The Role of Supercomputing in Sustainable Development

Supercomputers not only have the potential to reduce their own environmental impact, but they can also contribute to **global sustainability goals** in other ways. One of the most important areas where supercomputing can have a significant impact is in **climate modeling** and **environmental research**. By simulating complex climate systems and predicting future environmental scenarios, supercomputers help researchers develop strategies for mitigating the effects of climate change and transitioning to more sustainable practices.

For example, supercomputing is being used to model **carbon capture technologies**, optimize the **design of renewable energy systems**, and analyze the potential environmental effects of various policy decisions. Supercomputers are also integral to the **development of sustainable materials**—from biodegradable plastics to **solar cells** and **battery technologies**—by simulating the molecular behavior of materials and optimizing their properties.

In this way, the very technologies that pose an environmental challenge are also enabling solutions to address some of the most pressing environmental issues of our time.

Powering the Future Responsibly

The demand for more powerful supercomputers will only continue to grow as science and technology evolve, but it is essential that these systems are developed in a manner that minimizes their environmental impact. The supercomputing industry is already making strides toward greater **energy efficiency**, **sustainability**, and **carbon neutrality**, and as more innovations in **green computing** emerge, the future of supercomputing looks much more sustainable.

The challenges of powering supercomputers while reducing their environmental footprint are significant, but the solutions are already on the horizon. With continued research, collaboration, and investment in **energy-efficient hardware**, **cooling technologies**, and the use of **renewable energy**, supercomputing can continue to advance the frontiers of science and technology without unduly harming the planet.

By adopting responsible practices, the supercomputing industry can not only contribute to solving the world's most pressing problems but can also do so in a way that ensures a healthier, more sustainable future for generations to come.

CHAPTER 9

The Future of Supercomputing: Beyond the Exaflop Era

The era of exaflop-level supercomputers, represented by machines like **Frontier**, is already pushing the boundaries of computational power. These machines, capable of performing one quintillion calculations per second, have transformed scientific research, enabling breakthroughs across fields such as **climate modeling, genomics, artificial intelligence (AI)**, and **material science**. Yet, as impressive as exaflop performance is, it represents just a stepping stone toward the future of computing.

Looking ahead, we are poised to enter an era where **quantum computing, photonic computing**, and **new computational architectures** will take center stage. These next-generation technologies promise to accelerate computational power exponentially, far surpassing what even the most advanced supercomputers of today can achieve. As the field of supercomputing evolves, these innovations will reshape industries, accelerate scientific discovery, and present entirely new challenges in terms of architecture, energy consumption, and security.

This chapter explores the future trajectory of supercomputing, delving into **quantum computing**, **photonic computing**, and other emerging technologies that could redefine computational power. We will consider how these advances could lead to further breakthroughs, enabling the next generation of high-performance computing systems, which will drive new industries, enhance human potential, and help solve some of the most pressing problems facing humanity.

The Limits of Classical Supercomputing

Before delving into the exciting new frontiers of supercomputing, it is important to understand the limits of current **classical computing** systems. The progression from **petaflop** to **exaflop** systems like Frontier represents a significant leap forward, but it is clear that there are physical and practical limits to the performance improvements that can be achieved with traditional architectures.

For example, classical systems rely heavily on **silicon-based chips**, which are governed by the well-known **Moore's Law**. Moore's Law, which predicts that the number of transistors on a chip would double approximately every two years, has held true for several decades. However, as transistors continue to shrink to atomic scales, quantum effects such as tunneling and **leakage** become problematic, making it increasingly difficult to maintain performance improvements through traditional means.

Additionally, the **energy consumption** of large-scale supercomputers is becoming a critical limiting factor. While innovations in **energy-efficient hardware** and **cooling systems** have improved efficiency, the exponential growth in computational power required for scientific applications demands an ever-increasing amount of energy. This poses significant challenges, both in terms of sustainability and cost.

As a result, the field of supercomputing is transitioning toward **post-silicon** technologies that can overcome the limitations of classical computing. These next-generation computing paradigms promise to offer breakthroughs in both computational speed and energy efficiency, potentially unlocking an entirely new realm of scientific discovery and technological capability.

Quantum Computing: A Paradigm Shift in Supercomputing

Quantum computing is perhaps the most talked-about and hotly anticipated technology on the horizon for supercomputing. Unlike classical computers, which process information in **binary bits** (0s and 1s), quantum computers use **quantum bits** or **qubits**. Qubits can exist in multiple states simultaneously due to a phenomenon known as **superposition**, enabling quantum computers to perform many calculations at once. This property makes quantum computers vastly more powerful than classical systems for certain types of problems.

The potential applications of quantum computing are immense, especially in fields that require solving complex, multidimensional problems. Some key areas where quantum computing could significantly impact supercomputing include:

1. Cryptography: Quantum computers could potentially break traditional encryption methods, which rely on the difficulty of factoring large numbers. This could revolutionize cybersecurity and lead to the development of new cryptographic techniques that are resistant to quantum attacks.

2. Optimization: Quantum computing could provide dramatic speedups for optimization problems in logistics, finance, and artificial intelligence. Algorithms that take an impractical amount of time to solve on

classical computers could be solved in a fraction of the time using quantum techniques.

3. Material Science and Chemistry: Quantum computers could simulate the behavior of molecules and materials at a level of detail that is impossible with classical computers. This could lead to the discovery of new drugs, materials, and chemicals, accelerating the pace of innovation in industries such as pharmaceuticals and manufacturing.

4. Artificial Intelligence: Quantum machine learning could drastically speed up the training of large AI models, opening new possibilities for the development of autonomous systems and intelligent agents.

While quantum computing holds great promise, it is still in its infancy. The technology faces significant challenges, including **quantum decoherence** (loss of quantum information due to environmental interference), **error correction**, and the **scalability** of qubit systems. However, companies like **IBM**, **Google**, and **Honeywell**, as well as research institutions like **MIT** and **IBM Research**, are actively working to overcome these challenges and develop scalable, reliable quantum computers.

Photonic Computing: The Next Step in Speed and Efficiency

While quantum computing operates on the principles of quantum mechanics, **photonic computing** leverages the unique properties of **photons** (light particles) to process information. Photonic computing has the potential to offer several advantages over traditional electronic computing, including **higher speed, lower energy consumption**, and **increased bandwidth**.

Photonic computers use light to carry and process information, rather than relying on electrical signals. This enables **parallel processing** at much faster speeds than electronic circuits. Additionally, photons are less susceptible to **heat generation**, a key challenge in classical supercomputing systems that rely on electrical circuits. This means that photonic computers could operate with significantly less energy consumption and generate less heat, which would be a significant step forward in terms of sustainability.

Photonic computing is still in the experimental stage, but companies like **Intel, Microsoft**, and **Cisco** are exploring the potential of photonics to enhance computational power. One of the most exciting aspects of photonic computing is its potential to **integrate with quantum computing**, creating a hybrid system that combines the speed and efficiency of photonics with the computational power of quantum mechanics.

Neuromorphic Computing: Mimicking the Brain

Another promising area of research is **neuromorphic computing**, which seeks to emulate the architecture and functionality of the **human brain**. Unlike traditional computers, which operate on binary logic, neuromorphic systems are designed to mimic the brain's **neural networks**. This approach could enable computers to perform tasks such as pattern recognition, decision-making, and learning in ways that are more akin to human intelligence.

Neuromorphic computing relies on specialized hardware called **neuromorphic chips**, which are designed to replicate the behavior of biological neurons and synapses. These chips are able to process data in parallel and use **spiking neural networks** to carry out complex computations. By mimicking the way the brain processes information, neuromorphic systems have the potential to handle tasks such as **vision**, **speech recognition**, and **natural language processing** more efficiently than classical computers.

Neuromorphic computing could have a significant impact on areas such as **AI**, **machine learning**, and **robotics**, providing the computational power needed to develop more intelligent and adaptive systems. As AI continues to evolve, neuromorphic computing may be the key to creating systems that can learn, reason, and make decisions in a more human-like manner.

Hybrid Systems: The Future of Supercomputing

One of the most likely directions for the future of supercomputing is the development of **hybrid systems** that combine various computational paradigms. These systems will integrate classical computing with quantum, photonic, and neuromorphic technologies to create machines that are vastly more powerful and versatile than today's supercomputers.

A hybrid system might use classical supercomputing for general-purpose computations, quantum computing for complex problem-solving and simulation tasks, photonic computing for high-speed data processing, and neuromorphic computing for AI and machine learning applications. By combining the strengths of each technology, these hybrid systems could deliver unprecedented performance while addressing the challenges of energy consumption, scalability, and versatility.

The Next Breakthroughs: Beyond Exaflops

While exaflop-level supercomputers like **Frontier** represent a monumental achievement, they are merely the precursor to even greater breakthroughs in computational power. The future of supercomputing will likely be characterized by **multidimensional advancements** that span a variety of technologies, from **quantum** and **photonic computing** to **neuromorphic systems** and **hybrid architectures**. These innovations will drive the next wave of scientific discovery, enabling humanity to tackle the most complex problems facing society—ranging from **climate change** to **global health crises** and **space exploration**.

As we move beyond the exaflop era, the future of supercomputing will be marked by ever-increasing **collaborations between research institutions, tech companies, and governments**. The possibilities for innovation are endless, and as computational power grows, so too will our ability to solve problems that were once thought impossible.

The future of supercomputing is brighter than ever, with quantum, photonic, and neuromorphic computing opening up new avenues for technological progress. While we are currently in the midst of the exaflop era, the next decade will likely witness the arrival of systems that transcend these limitations, forever altering the landscape of science, technology, and human achievement.

CONCLUSION
The Role of Supercomputers in Solving Global Challenges

As we stand at the crossroads of technological evolution, the role of **supercomputers** in solving some of the world's most pressing challenges has never been clearer. From revolutionizing **scientific research** to shaping the future of **artificial intelligence**, **medicine**, and **sustainability**, supercomputers have become indispensable tools in our quest for a better, more equitable world. The breakthroughs that have already been made, and the immense potential for further advances, position supercomputing as a key driver in tackling some of the most difficult issues humanity faces in the 21st century.

In this book, we have explored the power and impact of **supercomputing** through a variety of lenses, from the technical innovations that power these systems, to their real-world applications across industries and scientific disciplines. As the global demand for advanced computational power continues to rise, supercomputers will play an increasingly critical role in solving complex, large-scale challenges such as **climate change**, **healthcare crises**,

food security, **energy transitions**, and **international security**. However, as powerful as these machines are, their capabilities also come with significant challenges that must be addressed, from **ethical considerations** and **environmental impacts** to the **future of computational architectures** and the need for **sustainable development**.

A Tool for Scientific Discovery

Supercomputers have already demonstrated their immense value in advancing our understanding of the natural world. With the ability to process vast datasets, run complex simulations, and model systems in ways that were previously inconceivable, these machines are transforming scientific disciplines across the board. From **climate modeling** to **genomics**, **material science**, and **drug development**, supercomputers enable researchers to simulate entire ecosystems, predict weather patterns with unprecedented accuracy, and design drugs and materials that can address global challenges.

For example, the ability to **simulate climate change** and model the Earth's changing atmosphere has made it possible to better understand the long-term implications of human activity on the planet. With more accurate climate models powered by supercomputers, scientists can predict shifts in global weather patterns, ocean currents, and ecosystems, offering crucial data for policymakers to take decisive action. Similarly, supercomputing's contribution to **genomic research** has revolutionized the study of genetics, speeding up the development of treatments for diseases such as cancer, Alzheimer's, and even the **global COVID-19 pandemic.**

In **material science**, supercomputers are enabling the development of new, innovative materials that can improve everything from electronics to clean energy technologies. These advancements are essential for solving global challenges such as transitioning to **renewable energy**, creating **energy-efficient infrastructure**, and tackling **the growing demand for sustainable resources**. The potential for **supercomputing** to drive breakthroughs in **advanced manufacturing, carbon capture,** and **energy storage** is profound and will play a significant role in our efforts to mitigate the impacts of **climate change.**

Advancements in Artificial Intelligence (AI) and Machine Learning

Perhaps the most transformative impact of supercomputing lies in its ability to accelerate progress in **artificial intelligence** and **machine learning.** As AI becomes increasingly central to solving complex global problems,

supercomputers provide the computational power necessary to train and deploy massive, complex models. Whether it's improving healthcare diagnostics, predicting financial trends, or optimizing supply chains, supercomputing enables AI systems to process and analyze vast amounts of data at speeds far beyond the capabilities of traditional computing.

For instance, AI-driven applications are making great strides in **predictive medicine**, where supercomputers are analyzing massive datasets of patient information to predict disease outbreaks, improve patient care, and develop personalized treatment plans. In the **agriculture** sector, AI algorithms, powered by supercomputing, are optimizing crop yields and reducing waste by predicting the optimal conditions for growing food in different parts of the world. In **transportation**, AI and machine learning models are being used to improve the safety, efficiency, and sustainability of transportation systems, including the development of **autonomous vehicles**.

Supercomputers also serve as essential tools in the research and development of **AI ethics**. As AI systems grow in sophistication, addressing issues such as **bias**, **privacy**, and **accountability** is critical. Supercomputers help test and validate AI models, ensuring that they are both efficient and fair. In a world where AI is increasingly embedded in decision-making processes—whether in healthcare, law enforcement, or social media—ensuring that these systems operate responsibly is essential to preventing unintended consequences and fostering public trust in technology.

The Role of Supercomputers in National Security and Global Stability

Beyond scientific discovery and technological innovation, supercomputers are also vital tools for **national security** and **global stability**. Governments and military agencies around the world leverage supercomputing to enhance **intelligence gathering**, model complex geopolitical scenarios, and develop cutting-edge defense technologies. In a world increasingly shaped by cybersecurity threats, supercomputers provide the necessary computing power to analyze large amounts of data for **threat detection** and **national defense**.

Supercomputers play an integral role in simulating **military strategies** and **warfare scenarios**, allowing nations to test various defense strategies without engaging in actual conflict. These simulations help policymakers anticipate potential threats and devise appropriate responses, thus reducing the likelihood of conflict. In addition, supercomputers are essential for **cryptography**—the science of securing communication and protecting sensitive information—ensuring the safety and integrity of global information systems.

However, the geopolitical implications of supercomputing cannot be ignored. As countries invest heavily in building the most powerful systems, competition for technological supremacy becomes a matter of national pride and strategic advantage. **Global security** could depend on the ability to develop, maintain, and safeguard these high-performance systems, which are vulnerable to cyberattacks and other forms of technological sabotage.

Environmental Sustainability and Green Computing

As the world's need for supercomputing power grows, the **environmental impact** of these machines has become an increasing concern. The energy demands of supercomputers are immense, and the carbon footprint of high-performance computing systems must be carefully managed to ensure that they do not contribute to the very problems they are working to solve. As the **climate crisis** intensifies, making supercomputing more **energy-efficient** and sustainable has become a critical area of focus.

Green computing technologies are already being developed to make supercomputing more energy-efficient. Innovations such as **liquid cooling systems**, **energy-efficient processors**, and **carbon-neutral data centers** are helping to reduce the environmental impact of high-performance computing. For example, some supercomputers now operate using **renewable energy** sources, such as solar or wind power, and are designed to operate in **carbon-neutral environments**.

The transition to more sustainable supercomputing practices is essential, not just for the future of high-performance computing, but for the global effort to combat climate change. By ensuring that these machines operate with minimal environmental impact, we can leverage their power to address the **climate crisis** while maintaining the integrity of the natural world. The development of **energy-efficient supercomputers** will be one of the key factors in driving the future of both **scientific research** and **sustainable development**.

The Road Ahead: The Future of Supercomputing

Looking to the future, the potential for supercomputing to solve global challenges is boundless. As we enter the **quantum computing** era, we may witness breakthroughs in computational power that render even today's exaflop machines obsolete. Quantum computers, capable of processing data in fundamentally new ways, could unlock solutions to problems that are currently beyond the reach of classical supercomputers, including those related to **material science**, **drug discovery**, and **climate modeling**.

Supercomputing will continue to evolve in **architecture** and **performance**, integrating new innovations in **AI**, **cloud computing**, and **edge computing** to further accelerate the pace of technological advancement. These next-generation systems will enable more accurate, more efficient simulations, improving our ability to predict and mitigate global challenges. The ability to model everything from **pandemics** to **global supply chains** to **energy grids** in real-time will be transformative, providing leaders and scientists with the tools they need to make informed decisions in the face of rapidly changing global circumstances.

However, with this growing power comes the need for responsibility. As we push the limits of supercomputing, we must also ensure that ethical concerns, such as **data privacy**, **security**, and **equitable access** to technology, are addressed. The future of supercomputing will depend not only on technological advancements but on how we manage and use this power to build a better, more sustainable world.

In summary, **supercomputing** has already demonstrated its transformative power in addressing global challenges, from advancing scientific research to improving healthcare, **combatting climate change**, and ensuring national security. As supercomputing continues to evolve, its potential to drive innovation, address complex problems, and improve the quality of life for people worldwide will only increase.

The future of supercomputing lies not just in creating faster, more powerful machines, but in harnessing that power to solve some of the most critical issues facing humanity. The role of **supercomputers** in **scientific discovery**, **AI development**, **sustainability**, and **global stability** will continue to grow, providing us with the tools we need to navigate the challenges of the future. As we embrace these new technologies, we must also be mindful of the ethical, environmental, and societal implications of their use.

The journey toward solving the world's most pressing challenges is long, but with the continued evolution of **supercomputing**, we can be hopeful that we are on the right path toward a brighter, more sustainable future for all.

www.ingramcontent.com/pod-product-compliance
Lightning Source LLC
Chambersburg PA
CBHW071039240526
45469CB00006BD/2271